Our Ever-Changing Environment

Shelly Buchanan, M.S.Ed.

Consultant

Caryn Williams, M.S.Ed.
Madison County Schools
Huntsville, AL

Image Credits: p.15 Bill Brooks/Alamy; p.29 (top) Charles Stirling/Alamy; p.27 (top) Florian Kopp/imagebroker/Alamy; p.28 (top) JGI/Jamie Grill/Blend Images/Alamy; p.11 (bottom) Manus Hüller/dieKleinert/Alamy; p.22 Terry Donnelly/Alamy; p.29 (bottom) Tim Pannell/Tetra Images/Alamy; p.26 (top) Hill Street StudiosE//Blend Images/age fotostock; p.21 Mint Frans Lanting/age fotostock; p.25 (bottom) Danita Delimont/Gallo Images/Getty Images; p.25 (top) Dietmar Heinz/Picture Press/Getty Images; p.8 Keystone-France/Getty Images; p.5 (top left), 20 (bottom), 24, 27 (bottom) iStock; p.16 National Park Service; p.9 (top) Beawiharta/Reuters/Newscom; p.18–19 (bottom) David Wall/DanitaDelimont/Newscom; pp.8–9 Geoff Renner/Robert Harding/Newscom; pp.26–27 John Boykin/Stock Connection Worldwide/Newscom; pp.10–11 Westend61/Newscom; p.20 (top) Theo Allofs/Minden Pictures/National Geographic Creative; p.17 (top) UNESCO; pp.6, 11 (top), 12–13 Wikimedia Commons; all other images from Shutterstock.

Teacher Created Materials

5301 Oceanus Drive
Huntington Beach, CA 92649-1030
http://www.tcmpub.com

ISBN 978-1-4333-7369-5

© 2015 Teacher Created Materials, Inc.
Made in China
Nordica.052015.CA21500197

Table of Contents

Changing Earth

Look at the world around you. What do you see? Maybe you see trees, birds, and blue skies. Maybe you see roads, cars, and buildings. Some of these things are natural, while others were made by humans. Earth did not always look the way it does today. Earth has been in the making for a very long time. Our planet has been changing for billions of years. Yes, *billions* of years—4.5 billion years to be exact!

Our planet started as a lumpy mass of rock. For millions of years, this was all there was. Over time, Earth has transformed, or changed. Today, our planet has massive mountains and deep canyons. There are long winding rivers and vast oceans. Natural forces continue to change Earth. Powerful earthquakes move the land, while violent **volcanoes erupt** to create new land. Rushing waters and gusty winds shape the land. All these forces combine with animals, plants, and people to transform Earth's surface.

Shifting and Shaking

The outer layer of Earth is called the *crust*. It is like the peel around an orange. But unlike an orange peel, Earth's crust is not one single piece. Instead, it is made up of giant pieces called *plates*. These plates can be dry land or the bottom of an ocean. The plates move too slowly for us to notice just by watching them. But over long periods of time, they can actually move mountains!

The arrows show how the plates move.

North American plate

Eurasian plate

Eurasian plate

Juan de Fuca plate

Caribbean plate

Arabian plate

Indian plate

Philippine plate

Cocos plate

Pacific plate

Nazca plate

South American plate

African plate

Australian plate

Scotica plate

Antarctic plate

Some plates move away from each other. Then water flows in to make lakes and oceans. Other plates move together to push up mountains and volcanoes. There are also plates that slide past each other. The process of Earth's plates moving is called **plate tectonics** (tek-TON-iks). This helps make the different kinds of land that we see on Earth. It is always changing Earth's surface.

Plate tectonics can create lakes and mountains.

Energetic Earthquakes

Have you ever felt an earthquake? The earth shakes all the time. Many times, people do not feel it. This is because, most of the time, the tectonic plates gently nudge and bump up against one another. But sometimes, pieces of crust get caught against each other. Pressure (PRESH-er) starts to build as these pieces try to slide by each other. Suddenly, the plates shift and slide quickly past each other. Then, there is a burst of movement as the pressure is released. These are the large earthquakes that we feel.

The land can change when an earthquake hits. Sometimes, large **valleys** are created. When rocks fall or land slides, streams can change course. Rivers can flow in a different direction! Earthquakes can damage things that we have built. Bridges break, buildings fall, and highways crumble. The results can be very dangerous for people living in the area.

These houses were destroyed by an earthquake in Chile in 1960.

Richter Scale

The Richter (RIK-ter) Scale is a measurement used for showing the strength of earthquakes. We usually do not feel quakes that measure 3.0 or less. The largest earthquake ever recorded happened in Chile in 1960. It measured 9.5!

A seismograph (SAHYZ-muh-graf) measures and records the movement of the earth during an earthquake.

This gap at a plate boundary in Iceland is quite wide.

Violent Volcanoes

There are more than a thousand active volcanoes on Earth. But today, we usually know when they will erupt.

Volcanoes erupt when **magma**, or hot liquid rock, pushes its way up to the surface. This happens when pressure builds deep below Earth's crust. The pressure pushes the magma up through a crack in the crust. Then, it comes out the top of a volcano.

Many eruptions (ih-RUHP-shuhnz) occur where the plates meet. Volcanoes may explode in a sudden burst. Or they may ooze and flow slowly downhill. Once out of the ground, the magma is called **lava**. It cools and creates new rocks. Other eruptions happen in hot spots. These are areas where it is so hot that it melts Earth's crust. The magma then blows through the surface. Many volcanoes erupt in our oceans. Some of these erupt with such force and so often that they become islands.

An undersea volcano erupts off the coast of the Solomon Islands.

ash

lava

magma

crust

Moving Mountains

Mountains are steep landforms. They can have sharp or rounded points that rise above the land around them. The highest point is called the *peak*, or summit. Some mountains stand alone. Others are part of a range, or group.

Mount Everest

Mount Everest is the tallest mountain in the world. It is located in Asia. It is a fold mountain. People have tried to climb to the top since 1921. In 1953, the first group finally made it! Today, many people climb Mount Everest every year.

There are many kinds of mountains. Each is made in a different way. Fold mountains are made when two plates are pushed together. Fault-block mountains form when land is pushed up on one side, and land sinks down on another side. Dome mountains are made when magma pushes up the crust. And volcanic mountains form when lava builds up on top of a volcano.

Mountains are found on land and in the sea. Many islands are really underwater mountains that reach above sea level. This is true of the Hawaiian (huh-WAHY-uhn) islands.

The Alps are fold mountains.

The Tetons are fault-block mountains.

Mt. Fuji is a volcanic mountain.

The Adirondacks are dome mountains.

Water World

Water is always moving around our planet. It flows in rivers and in oceans. It goes from the ocean into rain clouds that travel around the world. Rain falls from these clouds to water the land and help plants grow.

Waves can hit the land with great force.

These rocks will slowly be turned into sand.

Our Oceans

Ocean water is in constant motion. Waves break on the beach and stir up the water. Currents flow like rivers in the ocean. They move ocean water around the world. They are hard to see because they flow underwater. The moon and the sun's **gravity** create tides. This causes water levels to rise and fall.

Oceans change Earth's surface. Coastlines are shaped by ocean waves. The pounding surf hits the land. The rocks and pebbles in the water wear away when they hit the shoreline. They break down into sand, creating beaches.

This is high tide in New Brunswick, Canada.

This is low tide in New Brunswick, Canada.

Tidal Tug of War

Gravity is a force that pulls objects together. In a way, the moon and the sun play a game of tug-of-war with Earth's oceans. Their gravity pulls the ocean water towards themselves, causing water levels to rise and fall around the world.

Rushing Rivers

Rivers take water from mountains to the seas below. They water the land and flow back into the oceans. Rivers affect the land in many ways. They start out as rain or melting ice. The water flows downhill. Then, small streams join other streams to become rivers. They become larger as they move downhill toward the sea. Some rivers are wide and move slowly. Others are narrow and rush downhill in fierce rapids.

Rivers do important work for Earth's **environment** (en-VAHY-ruhn-muhnt). They bring needed water to the land. This is good for plants and animals that live there. It is also good for farmers growing crops. Plants need water to grow. Animals need water to live. Some plants and animals live in the rivers. We would not be able to survive if rivers did not bring water to us from the mountains.

These rivers flow into the Gulf of Mexico.

Missouri River

Mississppi River

Arkansas River

Ohio River

Gulf of Mexico

Why So Salty?

Have you ever tasted seawater? It is very salty. Rivers bring rocks to the sea. Over millions of years, the salt in these rocks made the sea salty.

The La Plata River in South America brings rocks and salt to the Atlantic Ocean.

the Mississippi River

Niagara (nahy-AG-ruh) Falls is located in both New York and Ontario, Canada.

Rivers bend and turn as they move around hard rock. They create rapids and waterfalls. As rivers flow across the land, they slowly wear it down. This is called **erosion** (ih-ROH-zhun).

Faster flowing water creates more erosion. Over time, rivers wear away dirt and rocks. After many years, a river can turn large rocks into sand. The sand and rock in the rivers rub against other rocks. This speeds up the erosion of the rocks. The river carries the sand to the ocean, making sandy beaches.

These arches and tunnels were eroded by the Oparara River in New Zealand.

The Colorado River eroded rock and made the Grand Canyon in Arizona.

Have you ever seen a deep canyon or steep valley? You may have noticed a river running through them. Water moving past rock slowly erodes it. After a very long time, the water cuts through the rock to form steep canyons and valleys. This is what created the Grand Canyon. Over time, water can change the way Earth looks.

A piece of a glacier falls into the ocean.

A glacier in Alaska flows into a bay.

Gigantic Glaciers

Glaciers (GLEY-sherz) are giant chunks of ice. They move very slowly because they are frozen. Most glaciers are found in Greenland and Antarctica (ant-AHRK-ti-kuh). These places are close to the north and south poles of Earth, where it is very cold.

Glaciers carve large land formations (fawr-MEY-shuhnz). This happens as they slowly scrape across the land. They can move dirt and rock hundreds of miles. Over long periods of time, they can make large valleys.

Many glaciers form in mountain valleys where there is a lot of snowfall. The fresh snow is pressed into layers of ice. This ice is heavy and starts to slide slowly down the valley. The glacier moves over rock and dirt. It erodes the land underneath and next to it. The glacier creates a deep valley as it travels. Finally, the glacier will reach a place where it is warmer and the ice will melt. Or, the glacier may flow into the ocean and break apart.

Glaciers carve out the land in Alaska.

Wind Power

What happens to your hair and clothes when you are outside on a windy day? What does the wind do to the leaves on the trees? The wind creates a lot of movement. It is strong enough to change the land. It slowly erodes rocks, and it moves sand and other materials to new places. Over time, this makes the land look different.

The wind is even more powerful in dry **desert** areas. In deserts there are few plants and very little water to hold the soil in place. The wind blows the soil and sand around. They rub against rocks and slowly erode them.

Sometimes, the wind creates **mesas**. These are large hills with flat tops and steep sides. The wind also forms **dunes**, or hills of sand. Sometimes, heavy winds start sandstorms that can change the landscape in just a few hours.

There are many sand dunes in Death Valley National Park in California.

This is Hunt's Mesa in Monument Valley in Utah.

Erosion helped form Arches National Park in Utah.

Our Living Earth

Living things on Earth must change as the planet changes. They need to adapt, or change, to live in the cold or heat, in deep oceans, or on tall mountains. At the same time, living things affect their environment, too. Our planet would not be the same without the things that live on it.

The roots of trees and plants hold soil in the ground. Some protect coastlines and riverbanks from erosion. The leaves of plants put **oxygen** (OK-si-juhn) in our atmosphere. In the ground, worms wiggle and move to bring air and water to the soil.

Larger animals make a difference, too. Beavers build dams to keep them safe from other animals. Dams are walls that stop or slow the flow of water. They may be made of stone, wood, or mud. Dams can stop rivers or change their direction. This can create ponds. Ponds provide a rich environment that allows many more plants and animals to live in that area.

This tree's roots hold the soil in place.

This beaver has long teeth to chew through wood.

Beavers created this dam in Grand Teton Park in Wyoming.

Humans have a huge effect on the environment. We use large amounts of Earth's **natural resources** (REE-sawr-suhz). We cut down trees and clear forests to farm the land and build homes. We construct bridges, towns, and cities. We put up dams so that we can use Earth's water to make power. We **mine** the land for minerals to make items such as computers and cars. We are forever making our environment adapt to keep up with the changes that we make.

Picking up trash makes our world a cleaner place.

Sometimes, we change our environment in bad ways. Today, there are fewer trees and more trash in the world. This hurts other living things. We need to replace the things from the planet that we use. We can plant new trees and pick up our trash. We can recycle. We need to keep Earth healthy so that it can keep us healthy. It is important for us to take care of our planet because our planet takes care of us.

This boy improves his community by planting trees.

This forest has been cut down.

Model It!

You have learned how wind, water, and plate tectonics all change Earth's surface. Now, teach someone else about it. Build a simple model that shows one of these forces in action. Maybe your model will show how a shoreline erodes or how wind creates a dune. Do your research. Be creative. And most importantly, have fun!

These kids make their model volcano erupt.

These boys create a pond to learn how water affects the environment.

These girls create a wind turbine model.

Glossary

desert—an area of very dry land with few plants and little rainfall

dunes—hills or ridges of sand piled up by the wind

environment—the natural world

erosion—the process of breaking something down by the action of water, wind, or glacial ice

erupt—to burst forth or to break through a surface

gravity—the natural force that causes things to move toward each other

lava—hot liquid rock above Earth's surface

magma—hot liquid rock below Earth's surface

mesas—hills that have flat tops and steep sides

mine—to dig in order to find and take away natural resources from the land

natural resources—things existing in the natural world that a country can use

oxygen—an element that is found in the air and is necessary for life

plate tectonics—the movements of large sections of Earth's surface

recycle—to make something new from something that has been used before

valleys—areas of low land between hills and mountains

volcanoes—mountains with holes in the tops or sides that sometimes send out rocks, ash, or lava in sudden explosions

Index

Your Turn!

Imagine That

Imagine that you are a scientist. You study Earth's ever-changing environment. You might study volcanoes, earthquakes, or erosion. Choose a topic from this book. Create a newspaper article that alerts people to how these things change Earth.